Unconventional Warfare

SINE PARI

Pocket Guide

The purpose of this document is to provide a pocket reference of Unconventional Warfare (UW) doctrine, concepts, academic inquiry, and suggested supplementary reading for military leaders and planners.

Joint Publication (JP) 3-05.1 defines UW as:
Activities conducted to enable a resistance movement or insurgency to coerce, disrupt or overthrow a government or occupying power by operating through or with an underground, auxiliary, and guerrilla force in a denied area.

Point of contact for this guide is the United States Army Special Operations Command, Deputy Chief of Staff G3, Sensitive Activities Division G3X, AOOP-SA, Fort Bragg, North Carolina 28310.

Contents

Introduction

This guide is a quick reference of Unconventional Warfare (UW) theory, principles, and tactics, techniques and procedures. It is not a complete treatment of the subject. To guide further study, it includes (in annotated form) as many references as possible starting with established law, policy and doctrine, includes scientific studies, and finishes with recommended reading on the subject.

The term UW often elicits strong responses both negative and positive, though many have a fundamental misunderstanding of the term itself, and its application supporting U.S. policy. Simply, UW is the support to a resistance movement. Historically and most often, the U.S. supported a semi-organized militarized irregular force -- known in doctrine as a Guerrilla Force, as part of an insurgency -- such as the U.S. support to the Afghanistan Northern Alliance in 2001. This is most often due to a foreign policy decision on when to get involved.

However, the application of UW is much broader and adaptive. The methods and techniques used and the planning for the operations are dependent on the state of the resistance movement, environment, and the desired end state. Support to a resistance organization in its incipient state requires significantly different planning and support than one in the war of movement state (using Mao's phases). The first will most likely be small, very sensitive, longer in duration, and often conducted under Title 50 authority, whereas the latter is large scale and open, as with the Northern Alliance in 2001.

Thus to prepare and train a force to conduct UW writ large, the theoretical construct must encompass all developmental states of the resistance; numerous environments, ideologies and circumstances; and account for all possible paths to a desired end state. The straw man and scenarios discussed in doctrine are not prescriptive but *descriptive* of a comprehensive campaign necessary to explore all possibilities, and their applicability.

UW has a wide range of applications in the contemporary environment, whether a textbook approached operation supporting the Syrian resistance, preparing a partner state ahead of potential occupation, or enabling a tribal group to resist Da'esh occupation in an Iraqi city. Special Operations Commanders must understand UW theories, principle, and tactics, and adapt them based on circumstance, the resistance, the opposition, and the desired end-state. This guide will help commanders and their staffs to find the relevant information necessary to understand and conduct UW.

Unconventional Warfare Overview

The focus in UW is on the indigenous resistance elements, not U.S. force structures and procedures. UW falls within the construct of Irregular Warfare (IW) and is one of U.S. Special Operations Command's (USSOCOM) Core Activities. USSOCOM Directives 10-1cc (U) and 525-89 (S//NF) establishes USASOC as the Lead Component for UW. This directive also describes the SOF service component capabilities and tasks for UW.

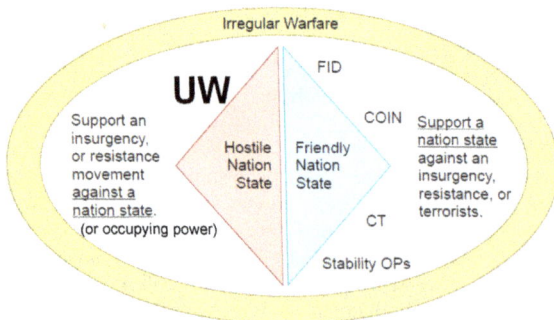

Figure 1. USSOCOM Activities

Special Operations Core Activities

- Unconventional Warfare (UW)
- Foreign Internal Defense (FID)
- Direct Action (DA)
- Civil Affairs Operations (CAO)
- Special Reconnaissance (SR)
- Security Force Assistance (SFA)
- Counterterrorism (CT)
- Hostage Rescue and Recovery (HRR)
- Counterinsurgency (COIN)
- Foreign Humanitarian Assistance (FHA)
- Countering Weapons of Mass Destruction (CWMD)
- Military Information Support Operations (MISO)

Terms Used in UW

Joint doctrine defines UW as activities conducted to enable a <u>resistance movement</u> or <u>insurgency</u> to <u>coerce</u>, <u>disrupt</u>, or <u>overthrow</u> a government or occupying power by operating through or with an <u>underground</u>, <u>auxiliary</u>, **and** <u>guerrilla force</u> in a denied area. (JP 3-05)

More recently, published Public Law defines UW as "activities conducted to enable a resistance movement or insurgency to coerce, disrupt, or overthrow a government or occupying power by operating through or with an underground, auxiliary, **or** guerrilla force in a denied area." [Public Law 114-92 Sec. 1097, S.1356 — 114th Congress (2015-2016), National Defense Authorization Act for FY 2016]

<u>Resistance Movement</u>: An organized effort by some portion of the civil population of a country to resist the legally established government or an occupying power and to disrupt civil order and stability. (JP 3-05)

<u>Insurgency</u>: The organized use of subversion and violence to seize, nullify, or challenge political control of a region. Insurgency can also refer to the group itself. (JP 3-24)

<u>Counter Insurgency</u>: Comprehensive civilian and military efforts taken to defeat an insurgency and to address any core grievances. (JP 3-24)

<u>Foreign Internal Defense</u>: Participation by civilian and military agencies of a government in any of the action programs taken by another government or other designated organization to free and protect its society from subversion, lawlessness, insurgency, terrorism, and other threats to its security. (JP 3-22)

<u>Coerce</u>: Coercion is forcing someone of some entity to do something it would rather not do. UW can apply the method of coercion through supporting a resistance or insurgency. (JP 3-05.1)

<u>Disrupt</u>: Disruption prevents or impedes someone or some entity from doing something it would prefer to do. Although disruption can be a relatively small-scale, it can also be large-scale such as coordinated regional resistance. (JP 3-05.1)

Overthrow: The USG may sponsor UW to overthrow a state or occupying power when it is intended that the supported successful resistance will support appropriate leaders for political control and governance. (JP 3-05.1)

Underground: Cellular organization within the resistance that has the ability to conduct operations in areas that are inaccessible to guerrillas, such as urban areas under the control of the local security forces. (JP 3-05.1)

Auxiliary: Refers to that portion of the population that provides active clandestine support to the guerrilla force or the underground. (JP 3-05.1)

Guerrilla Force: A group of irregular, predominantly indigenous personnel organized along military lines to conduct military and paramilitary operations in enemy-held, hostile, or denied territory. (JP 3-05)

Denied Area: An area under enemy or unfriendly control in which friendly forces cannot expect to operate successfully within existing operational constraints and force capabilities. (JP 3-05)

Components of a Resistance

Indigenous populations engaged in resistance are composed of the following primary components: the underground, auxiliary, guerrilla forces, public components, shadow government, and a government-in-exile. The goals, objectives, and success of the resistance will determine the level of development and relationships among the components.

- The underground and guerrillas are politico-military entities that may conduct both political and military acts, and which represent the ends of a spectrum between clandestine and overt resistance.

- The auxiliary represents a clandestine support structure for both the underground and guerrillas.

- The public component functions as an overt, political, and/or material support entity. The public components may negotiate with the nation-state government or occupying power on behalf of resistance movement objectives, and will typically make overt appeals for domestic and international support. Public components may represent resistance strategic leadership or merely an interest section.

The underground is a cellular organization within the resistance that has the ability to conduct operations in areas that are inaccessible to guerrillas, such as urban areas under the control of the local security forces. Examples of underground functions include: intelligence, counterintelligence (CI) networks, special material fabrication (example: false identification), munitions, subversive radio, media networks (newspaper or leaflet print shops), social media, webpages, logistic networks, sabotage, clandestine medical facilities, and generation of funding.

The auxiliary refers to that portion of the population that provides active clandestine support to the guerrilla force or the underground. Members of the auxiliary are part-time volunteers who have value because of their normal position in the community. Some functions include: logistics procurement and distribution, labor for special materials, early warning for underground facilities and guerrilla bases, intelligence collection recruitment, communications couriers or messengers, distribution media and safe house management.

A guerrilla is an irregular, predominantly indigenous member of a guerrilla force, organized similarly to military concepts and structure to conduct military and paramilitary operations in enemy held, hostile, or denied territory. Guerrillas are neither militias nor mercenary soldiers whose allegiance is secured solely by payment, nor are they criminal gangs conducting illegal activities for profit.

The public component refers to an overt political manifestation of a resistance. Public components are primarily responsible for negotiations with the nation-state government or occupying power representatives on behalf of resistance movement objectives. Every case of resistance is unique, and the degree to which public representatives exist will vary. If the nation-state decides to suppress these public components completely, the public components may have to dissolve and go underground. The public component is not synonymous with shadow government or government-in-exile.

A shadow government consists of governmental elements and activities performed by an irregular organization that replaces the governance functions of the existing regime (examples include security, health services, and taxation). Members of the shadow government can originate from any element of the irregular organization (e.g., underground, auxiliary, or guerrilla force). The shadow government operates in the denied area of an occupied territory.

A government-in-exile is a government displaced from its country of origin, yet remains recognized as a legitimate sovereign authority of a nation. A government-in-exile will normally take up sanctuary in a nearby allied or friendly nation-state.

Figure 2 below depicts the typical activities conducted by an insurgency or resistance movement. Most or all of the represented activities will be conducted by a successful resistance or insurgency regardless of whether the USG makes the policy decision to support it. USG support to resistance – UW – is initiated by either a presidential finding or an EXORD. UW then proceeds generally following a 7-phase conceptual model called the Seven Phases of Unconventional Warfare.

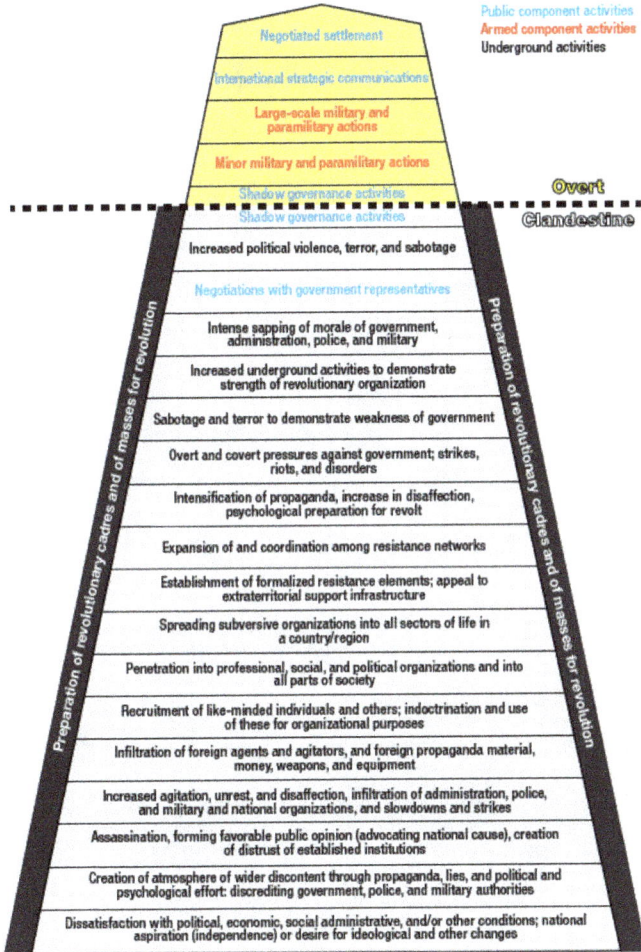

Figure 2. Activities of an Insurgency or Resistance Movement

Doctrinal Template for UW

Army doctrine describes ARSOF's role in UW as phases. ARSOF activities before a presidential finding include shaping or conducting preparation of the environment. The extent of these activities is dependent on Global Combatant Command requirements and ARSOF access and placement. Other DOD terms used to describe this period include *steady state*, *left of bang*, and *persistent engagement*. This document describes this period of time as Phase 0. Since each UW campaign is unique, all phases of UW may not occur, depending upon U.S. objectives, the scale and tempo of the campaign, and the successes and failures of the resistance. Additionally, not all geographic areas where the resistance is active may be in the same phase during a UW campaign.

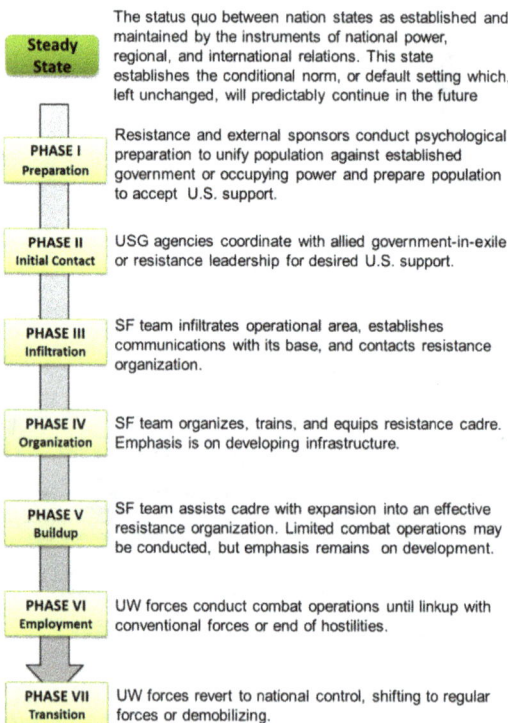

Steady State	The status quo between nation states as established and maintained by the instruments of national power, regional, and international relations. This state establishes the conditional norm, or default setting which, left unchanged, will predictably continue in the future
PHASE I Preparation	Resistance and external sponsors conduct psychological preparation to unify population against established government or occupying power and prepare population to accept U.S. support.
PHASE II Initial Contact	USG agencies coordinate with allied government-in-exile or resistance leadership for desired U.S. support.
PHASE III Infiltration	SF team infiltrates operational area, establishes communications with its base, and contacts resistance organization.
PHASE IV Organization	SF team organizes, trains, and equips resistance cadre. Emphasis is on developing infrastructure.
PHASE V Buildup	SF team assists cadre with expansion into an effective resistance organization. Limited combat operations may be conducted, but emphasis remains on development.
PHASE VI Employment	UW forces conduct combat operations until linkup with conventional forces or end of hostilities.
PHASE VII Transition	UW forces revert to national control, shifting to regular forces or demobilizing.

Figure 4. Phase of UW

Phase 0: Steady State

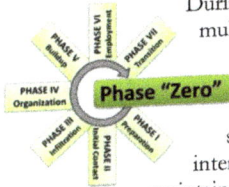

During the Steady State, the USG conducts Joint or multinational operations or interagency activities to dissuade or deter potential adversaries and to assure or solidify relationships with friends and allies. These activities (below) normally precede, sometimes by years, the operations they are intended to support. They are critical to establish, maintain, and reestablish the conditions whereby UW could be considered as a feasible U.S. strategic option. SOF can conduct Phase 0 activities continuously and in all operational modes (overt, low visibility, clandestine, and covert), and can include the full menu of theater cooperation engagement activities, and all preparation of the environment (PE) activities.

Activities
- Conduct continual Area Assessments
- Conduct Preparation of the Environment (PE): Civil, Military, Physical, Virtual Domains
- Identify Threats, and Design and Plan UW Options
- Activities to Legitimize Narratives Supporting U.S. Interests and Potential Resistance Movements
- Build Internal and External Support for Potential Resistance Movements
- Conduct Activities to Set Conditions for the Introduction of U.S. Forces into the Area of Operations When Necessary
- Conduct Continual Military Information Support Operations (MISO) Assessments
- Maintain PR Intelligence Analysis to Support Assisted Recovery Mechanisms

UW Phase I: Preparation

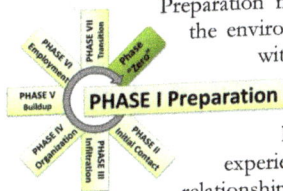

Preparation must begin with intelligence preparation of the environment (IPOE) to understand the dynamics within the populace. UW IPOE includes, but is not limited to, a thorough analysis of the resistance force's strengths, weaknesses, logistics concerns, level of training and experience, political or military agendas, factional relationships, and external political ties. Along with this data, IOPE requires a thorough target area study. At a minimum, the target area study includes governmental services, living conditions, and political, religious, economic, environmental, medical, and educational issues.

11

Analysis and planning for transition to civil governance begins, to include contingency planning for collapse of the adversary government or power and sudden victory by the resistance.

Activities
- Conduct Continual Area Assessment
- Design, Plan, and Update the UW Campaign
- Gain Access to and Identify Resistance Assets
- Continue Preparation of the Environment (PE): Civil, Military, Physical, Virtual Domains
- Conduct MISO and Civil Affairs (CA) Support to UW
- Conduct Joint Intelligence Preparation of the Operational Environment (JIPOE)
- Synchronize Activities with Interagency Partners
- Develop Non-Standard Logistics (NSL) Infrastructure and Plan Support to Irregular Forces
- Maintain Support Mechanisms
- Maintain/Develop PR Intelligence Analysis to Support Nonconventional Assisted Recovery (NAR)/Unconventional Assisted Recovery (UAR) mechanisms

UW Phase II: Initial Contact

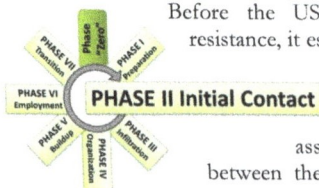

Before the USG decides to render support to a resistance, it establishes contact with representatives of a resistance organization to assess the compatibility of U.S. and resistance interests and objectives. This assessment is largely a political negotiation between the USG and the resistance organization.

Once the USG establishes compatibility, it assesses the resistance potential. During the initial contact, planners may arrange for the dispatch and reception of a pilot team. When possible, planners exfiltrate a resistance representative, referred to as an asset, from the operational area to brief the pilot team during its planning phase. The asset may accompany the team during their infiltration into the operational area and facilitate their linkup with resistance forces.

Activities
- Determine Mission Command Requirements
- Establish Contact with Resistance
- Prepare Pilot Teams and Synchronize with IA Partners
- Continue PE: Civil, Military, Physical, Virtual Domains

- Conduct MISO and CA Support to UW
- Conduct JIPOE and continue R&S
- Synchronize Activities with IA Partners
- Establish Non-Standard Logistics (NSL) Stocks and Build Capacity to Support Irregular Forces
- Continue PR Analysis in Support of PR Mechanisms

UW Phase III: Infiltration

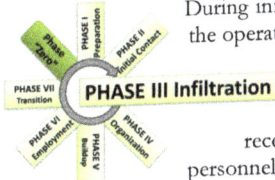

During infiltration, a pilot team clandestinely infiltrates the operational area to link up with the resistance force and to conduct or confirm a feasibility assessment. If the assessment is favorable, the pilot team coordinates to infiltrate and receive follow-on SF teams and supplies. MISO personnel, attached to follow-on SF teams, provide SFODAs with an early influence and information capability while developing an indigenous capability. As teams infiltrate the operational area and link up with their respective resistance-force counterparts, they begin their own operational assessment to confirm or deny the assumptions of the overarching UW campaign plan.

Activities
- Infiltrate Pilot Teams
- Conduct UW Area Assessment
- Establish Mission Command Infrastructure
- Minimize Risks to Force/Mission
- Continue PE: Civil, Military, Physical, Virtual Domains
- Conduct MISO and CA Support to UW
- Conduct JIPOE and Continue R&S
- Synchronize SOF, Resistance, and Interagency Activities
- Establish NSL Stocks and Plan Support to Irregular Forces
- Maintain Support Mechanisms
- Expand PR Infrastructure

UW Phase IV: Organization

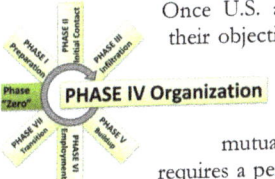

Once U.S. advisors link up with resistance leadership, their objective is to determine and agree upon a plan to organize the resistance for expanded operations. In addition to physical preparations, this entails a confirmation of mutual objectives and prior agreements. This requires a period of rapport-building to develop trust and

13

confidence, as well as a period of discussion of expectations from both sides. Before a resistance organization can successfully engage in combat operations, its leadership must organize an infrastructure that can sustain itself in combat and withstand the anticipated hostile reaction to armed resistance. During organization, the resistance leadership develops a resistance cadre as the organizational nucleus for the buildup phase.

Activities
- Develop Resistance Campaign Plans
- Establish Operations and Intel Infrastructure
- Organize and Train Resistance Elements (Underground, Auxiliary, Guerrilla Forces)
- Enhance Force Protection Measures (FORCEPRO) and Counter-Intelligence Capabilities
- Refine MISO Messaging
- Develop Area Complex and Logistics Infrastructure
- Expand Sanctuaries and Safe Havens
- Intensify UW Activities with OGAs
- Shadow Government Organizes Civil Support and Services
- Continue to Build PR (NAR/UAR) Architecture

UW Phase V: Buildup

During buildup, the resistance cadre improves the organization's clandestine supporting infrastructure to prepare for expanded offensive operations. When the organization begins to conduct operations of wider scope and across a wider area, many of these operations will draw attention from counter-guerrilla forces. The organization must have the supporting clandestine infrastructure to prepare for and sustain these operations.

Activities
- Grow Resistance Organization:
  - Integrate Disparate Resistance Groups
  - Increase Recruitment
  - Intensify/Advance Training
- Enhance Targeting Capabilities, and Expand Operational Reach and Effectiveness
- Expand Persistent and Intrusive Reconnaissance and Surveillance (R&S) Operations
- Intensify MISO Messaging

14

- Synchronize Operational Effects with Joint Task Forces, other U.S. Government Agencies, and Coalition Partners
- Continue to Build PR (NAR/UAR) Architecture
- Expand Logistics Stocks, Support, and Sustainment Capacity

UW Phase VI: Employment

During employment, the resistance force initiates an expanded scope of offensive operations to achieve the desired effects. The desired effects can range from causing the enemy to commit limited resources away from a pending invasion area, support to a pending invasion area (as in general war), or in the case of a separate insurgency (limited war), the achievement of specific strategic politico-military objectives. The main activities in this phase consist of interdiction and MISO. The specific tactics (raids and ambushes) and activities (intelligence gathering, force protection) that occur during employment are not exclusive to UW.

Activities
- Maintain Alignment of Resistance Campaign Activities with Resistance Narratives and Legitimacy
- Synchronize Resistance Operations and Activities Across Boundaries
- Expand Resistance Controlled Territory
- Employ MISO Messaging to Delegitimize Adversary and Legitimize Resistance
- Provide Civil-Military Support to Displaced Persons, Refugees & Evacuees (DPRE), and Expand and Prepare Resistance Capacity for Governance

UW Phase VII: Transition

SOF may conduct UW until they assist the resistance in removing the hostile power and the indigenous population becomes the government. At this point, it is critical to shift mindsets from defeating the adversary regime to protecting the newly installed government and its security personnel from insurgency, lawlessness, and subversion by former regime elements that attempt to organize their own resistance. Planners should have addressed transition planning in the feasibility assessment that formed the basis of the determination to support the resistance organization.

15

Activities

- Promote New Government Legitimacy
- Transition from SOJTF to Security Forces Assistance (SFA)
- Demobilize and Integrate Guerrilla Forces, Underground and Auxiliary
- Provide Civil-Military Operations (CMO) Support to New Government
- MISO Themes Promote and Reinforce Civil Governance
- Be Prepared to Conduct FID/COIN
- Shadow Government and Government-in-Exile Transition Responsibility for Civil Governance

Coordinating Foreign Disclosure and Release of U.S. TTPs

The sharing and release of information to a resistance or other partners is part of the planning process for UW campaigns. Training, advising, and accompanying elements of a resistance will expose US TTPs, including sensitive and protected methods, to foreign nationals, and the effects of providing or withholding this information should be a deliberate decision.

Release, Transfer, and Export of Special Operations Tactics, Techniques, and Procedures. The release of classified military information (CMI) is governed by U.S. Law. Official government unclassified information must also be reviewed before disclosure to the public. Requests for information from the public, citing the Freedom of Information Act (FOIA), are processed through FOIA channels. Requests for information from foreign governments through government channels are processed by the USSOCOM Foreign Disclosure Officer (FDO).

Office of Primary Responsibility (OPR). The OPR for Unconventional Warfare (UW) is the USSOCOM J3X UW/PE Branch. Release of information regarding the tactics, techniques, and procedures (TTP), and doctrine for Special Operations (SO) capabilities associated with the conduct of UW is prohibited without the written consent of the USSOCOM OPR.

Request for release, transfer and export of SO TTP's will comply with USSOCOM Directive 350-27. The International Programs Branch (SCSO J32-IP) is the USSOCOM OPR for SO TTP Transfers. Proposed transfers are considered on a case-by-case- basis, in consultation with the UW OPR, and supported by USSOCOM when the transfer results in tangible and direct benefit to U.S. foreign policy and national security objectives. Request for release, extraction or dissemination of UW or associated TTP's are processed through the UW OPR.

RESOURCES

Law and Policy

Title 10, U.S. Code, Armed Forces
The baseline US legal authority for UW is in Title 10, USC, Section 167(j). This provision states that UW is one of the activities of USSOCOM.
http://uscode.house.gov/browse/prelim@title10&edition=prelim

Title 22, U.S. Code, Foreign Relations and Intercourse
Essential to understand Chief of Mission authorities to design, plan, and execute UW campaigns and PE activities.
http://uscode.house.gov/browse/prelim@title22&edition=prelim

Title 50, U.S. Code, War and National Defense
Provides presidential authority and limitations in regard to authorizing covert action. Requires the congressional intelligence committees to be kept fully and currently informed of all covert actions which are the responsibility of, are engaged in by, or are carried out for or on behalf of, any department, agency, or entity of the USG.
http://uscode.house.gov/browse/prelim@title50&edition=prelim

Public Law 114-92, Section 1097
Provides a more expansive definition of Unconventional Warfare (UW) that allows activities conducted with an underground, auxiliary, or guerrilla force to meet the threshold for UW.

Executive Order 12333, United States Intelligence Activities, Signed 4 December 1984 [As amended by Executive Orders 13284 (2003), 13355 (2004) and 13470 (2008)]
Lays out roles for various intelligence agencies and provides guidelines for actions of intelligence agencies.
www.archives.gov/federal-register/codification/executive-order/12333.html

Executive Order 13470, 30 July 2008
Amended EO 12333 to strengthen the role of the Director of National Intelligence.
www.gpo.gov/fdsys/pkg/FR-2008-08-04/pdf/E8-17940.pdf

DoDD 5100.01, Functions of the Department of Defense and Its Major Components, 21 December 2010
Describes organizational relationships within DoD, delineates major functions, supports DoD strategic processes and aligns with the overall goals and priorities of DoD.
www.dtic.mil/whs/directives/corres/pdf/510001p.pdf

DoDD 3000.07, Irregular Warfare, 28 August 2014
Establishes policy and assigns responsibilities for DoD conduct of IW and development of capabilities to address irregular challenges or threats to national security. IW includes relevant

DoD activities and operations such as CT, UW, FID, COIN, and stability operations that involve establishing or re-establishing order in a fragile state or territory.
www.dtic.mil/whs/directives/corres/pdf/300007p.pdf

DODI 3002.04, DOD Personnel Recovery—Non-Conventional Assisted Recovery (NAR), 17 November 2014
Establishes policy and assigns responsibilities for NAR as an integral part of personnel recovery in the DoD; Prescribes procedures for the development, planning, and integration of NAR in cooperation with intelligence, special operations, and interagency partners.
www.dtic.mil/whs/directives/corres/pdf/300007p.pdf

USSOCOM Directive 525-5 (S//NF) Advanced Special Operations Techniques (U), 14 November 2013
Provides guidance and assigns responsibilities for the conduct of activities using Advance Special Operations Techniques (ASOT) by SOF.
(SIPRNET) *https://intelshare.intelink.sgov.gov/sites/usasocg3x/UWpedia/UWLibrary*

USSOCOM Directive 525-16 (S//NF) Preparation of the Environment (U), 14 November 2013
Establishes common concepts and provides general guidance for the planning, conduct, and resourcing of PE by SOF. PE includes those actions taken to prepare the operational environment for potential operations.
(SIPRNET) *https://intelshare.intelink.sgov.gov/sites/usasocg3x/UWpedia/UWLibrary*

USSOCOM Directive 525-89 (S//NF) Unconventional Warfare (U), 31 May 2012
Establishes USASOC as Lead Component for Unconventional Warfare (UW).
(SIPRNET) *https://intelshare.intelink.sgov.gov/sites/usasocg3x/UWpedia/UWLibrary*

Executive Order, Hostage Recovery Activities, 24 June 2015
Establishes a single USG operational body to coordinate all efforts for the recovery of U.S. nationals taken hostage abroad, with policy guidance coordinated through the National Security Council.
https://www.whitehouse.gov/the-press-office/2015/06/24/executive-order-and-presidential-policy-directive-hostage-recovery

Presidential Policy Directive/PPD-29 U.S. National Taken Hostage Abroad and Personnel Recovery Efforts, 24 June 2015
PPD-29, including its classified annex, supersedes and revokes NSPD-12. This policy directs a renewed, more agile USG response to hostage taking of U.S. nationals and other specified individuals abroad.
https://www.whitehouse.gov/the-press-office/2015/06/24/presidential-policy-directive-hostage-recovery-activities

Doctrine: Joint, Army, and ARSOF Pubs

Joint Publications (JP)

JP 3-05 Special Operations, 16 July 2014
Provides overarching doctrine for special operations and the employment and support for special operations forces (SOF) across the range of military operations. Sets forth joint doctrine to govern the activities and performance of the Armed Forces of the United States in joint operations and provides the doctrinal basis for interagency coordination and involvement in multinational operations. Provides an overview for special operations and describes special operations core activities. Describes command and control of SOF and discusses the support considerations for SOF, provides military guidance to combatant commands and joint force commanders, and prescribes joint doctrine for operations, education, and training.
http://www.dtic.mil/doctrine/new_pubs/jp3_05.pdf

JP 3-05.1, Unconventional Warfare, 15 September 2015
This publication has been prepared under the direction of the CJCS. Sets forth joint doctrine to govern the activities and performance of the Armed Forces of the United States in joint operations, and it provides considerations for military interaction with governmental and nongovernmental agencies, multinational forces, and other interorganizational partners. Provides joint doctrine to assess, plan, and execute unconventional warfare.
https://jdeis.js.mil/jdeis/new_pubs/jp3_05_1.pdf

JP 3-50 Personnel Recovery, 02 October 2015
This publication has been prepared under the direction of the CJCS. Sets forth joint doctrine to govern the activities and performance of the Armed Forces of the United States in joint operations, and it provides considerations for military interaction with governmental and nongovernmental agencies, multinational forces, and other interorganizational partners. Provides doctrine for the preparation, planning, execution, and assessment of personnel recovery.
https://jdeis.js.mil/jdeis/new_pubs/jp3_50.pdf

JP 3-50 APP B, (S//NF) Personnel Recovery-Appendix B, Classified Intelligence Support to Personnel Recovery (U), 02 October 2015
(SIPR ACCESS ONLY)

JP 3-50 APP E, (S//NF) Personnel Recovery-Appendix E, Classified Planning Supplement (U), 02 October 2015
(SIPR ACCESS ONLY)

Universal Joint Task List (UJTL)

Universal Joint Task List (UJTL), 15 March 2016
The Universal Joint Task List (UJTL) is a menu of tasks in a common language, which serves as the foundation for joint operations planning across the range of military and interagency operations. The UJTL supports DOD to conduct joint force development, readiness reporting,

experimentation, joint training and education, and lessons learned. It is the basic language in developing joint mission essential task lists (JMETL) and agency mission essential task lists (AMETL).

http://www.dtic.mil/doctrine/training/ujtl_tasks.pdf

SN 3.8 Conduct Special Operations Activities, 06 January 2015

Conduct full-spectrum special operations activities to support or achieve national strategic objectives. JP 3-05 (primary). This task includes establishing strategic appreciation and conducting mission analysis prior to execution. The Joint Staff / Secretary of Defense (SecDef) may designate Commander, United States Special Operations Command (CDRUSSOCOM) as supported commander for execution of global operations, including special operations activities against terrorists and their networks. Special operations core activities include: direct action (DA), special reconnaissance (SR), countering weapons of mass destruction (CWMD), counterterrorism (CT), unconventional warfare (UW), foreign internal defense (FID), security force assistance (SFA), hostage rescue and recovery, counterinsurgency (COIN), foreign humanitarian assistance (FHA), military information support operations (MISO), civil affairs (CA) operations, and other activities as directed by the President or the SecDef.

http://www.dtic.mil/doctrine/training/ujtl_tasks.pdf

ST 1.3.7 Conduct Unconventional Warfare (UW), Change 01 February 2016

Conduct military and paramilitary operations, normally of long duration, to enable a resistance movement or insurgency to coerce, disrupt, or overthrow a government or occupying power by operating through or with an underground, auxiliary, and guerrilla force in a denied area. JP 3-05, JP 3-05.1 (primary), CJCSI 3126.01A, CJCSI 3210.06A, DODD 3000.07. This task can be performed across joint operations areas (JOAs), and integrates and synchronizes indigenous or surrogate forces that are organized, trained, equipped, supported, and directed by an external source. It includes direct offensive, low visibility, covert or clandestine operations, as well as indirect activities of subversion, sabotage, intelligence activities, and evasion and escape. This task may require language proficiency and/or regional expertise and cultural knowledge to effectively communicate with and/or understand the cultures of coalition forces, international partners, and/or local populations and/or understand the operational environment (OE).

OP 1.2.4.8 Conduct Unconventional Warfare (UW), 06 January 2015

Enable a resistance movement or insurgency to coerce, disrupt, or overthrow a government or occupying power by operating through or with an underground, auxiliary, and guerrilla force in a denied area. JP 3-05 (primary), JP 3-05.1, CJCSI 3126.01A, CJCSI 3210.06, DODD 3000.07. The paramilitary aspect of unconventional warfare (UW) may place the Department of Defense (DOD) in a supporting role to interorganizational partners. The necessity to operate with a varying mix of clandestine/covert means and ways places a premium on operations security and actionable intelligence. This task may require language proficiency and/or regional expertise and cultural knowledge to effectively communicate with and/or understand the cultures of coalition forces, international partners, and/or local populations and/or understand the operational environment (OE).

Army Doctrine and Training Publications (ADRP)

ADRP 1-03, The Army Universal Task List (AUTL), 02 October 2015
ADRP 1-03 provides the structure and content of the Army Universal Task List (AUTL). The AUTL is intended to inform all members of the Profession of Arms of what the Army contributes to the joint force in terms of tasks performed. Additionally, it is intended that proponent training developers use the AUTL to develop more comprehensive training and evaluation outline evaluation criteria for collective tasks and proponent combat developers to better understand the tasks a given unit must perform. The Army Universal Task List (AUTL) describes what well-trained, well-led, and well-equipped Soldiers do for the Nation. While focused on the land dimension, abilities of Army forces complement abilities of other Services. The ability of Army forces to perform tasks builds the credible land power necessary for joint force commanders to preclude and deter enemy action, win decisively if deterrence fails, and establish a rapid return to sustained stability.
http://armypubs.army.mil/doctrine/DR_pubs/dr_a/pdf/adrp1_03.pdf

ADP 3-05, Special Operations, 31 August 2012
Describes the role of United States Army Special Operations Forces (ARSOF) in the U.S. Army's operating concept to shape operational environments in the countries and regions of consequence, prevent conflict through the application of special operations and conventional deterrence, and when necessary help win our nation's wars. Outlines ARSOF's requirement to provide in the nation's defense unequalled surgical strike and special warfare capabilities. Together these two different but mutually supporting forms of special operations comprise the American Way of Special Operations Warfighting. Provides a broad understanding of Army special operations by describing how executing the two mutually supporting critical capabilities of special warfare and surgical strike contribute to unified land operations. Provides a foundation for how the Army meets the joint force commander's needs by appropriately blending ARSOF and conventional forces. Defines and discusses special operations in the strategic context within which ARSOF expect to operate. It also discusses the roles and critical capabilities of Army special operations and describes the principles, regional mechanisms, characteristics, and imperatives of ARSOF. The principal audience of commanders and staffs of Army headquarters serving as joint task force or multinational headquarters should also refer to applicable joint or multinational doctrine concerning the range of military operations and joint or multinational forces.
http://armypubs.army.mil/doctrine/DR_pubs/dr_a/pdf/adp3_05.pdf

ADRP 3-05, Special Operations, 31 August 2012
Army Doctrine Reference Publication (ADRP) 3-05, Special Operations, provides a broad understanding of Army special operations by describing how executing the two mutually supporting critical capabilities of special warfare and surgical strike contribute to unified land operations. ADRP 3-05 provides a foundation for how the Army meets the joint force commander's needs by appropriate integration of Army special operations forces (ARSOF) and conventional forces. Army special operations forces are those Active and Reserve Component Army forces designated by the Secretary of Defense that are specifically organized, trained, and equipped to conduct and support special operations. The acronym ARSOF represents Civil Affairs (CA), Military Information Support operations (MISO), Rangers, Special Forces (SF), Special Mission Units, and Army special operations aviation forces assigned to the United States Army Special

Operations Command (USASOC)—all supported by the Sustainment Brigade (Special Operations) (Airborne) (SB[SO][A]).

http://armypubs.army.mil/doctrine/DR_pubs/dr_a/pdf/adrp3_05.pdf

Army Tactical Tasks (ART)

ART 7.6.4 Conduct Irregular Warfare

Irregular warfare is a violent struggle among state and non-state actors for legitimacy and influence over the relevant population(s) (JP 1). United States Army forces operations grouped under irregular warfare are foreign internal defense, support for insurgencies, counterinsurgency, combating terrorism, and unconventional warfare. (JP 3-26) (JS)

https://rdl.train.army.mil/catalog-ws/view/100.ATSC/60E015C4-01FE-4235-8775-F386C48654A7-1361576471062/report.pdf

ART 7.6.4.5 Conduct Unconventional Warfare (UW)

Unconventional warfare is a broad spectrum of military and paramilitary operations, normally of long duration, predominantly conducted through, with, or by host-nation or surrogate forces that are organized, trained, equipped, supported, and directed in varying degrees by an external source. It includes, but is not limited to, guerrilla warfare, subversion, sabotage, intelligence activities, and unconventional assisted recovery. Unconventional warfare is operations conducted by, with, or through irregular forces in support of a resistance movement, insurgency, or conventional military operations. (ATP 3-05.1) (USAJFKSWCS)

https://rdl.train.army.mil/

Field Manuals (FM)

FM 3-05 Army Special Operations, January 2014

Field Manual 3-05, Army Special Operations, provides the United States (U.S.) Army special operations forces commander and staff information on the structure and functions involved in Army special operations forces activities. Army special operations forces represent all forces assigned to the United States Army Special Operations Command: Special Forces, Military Information Support Operations, Civil Affairs, Rangers, Army special operations aviation, the 528th Sustainment Brigade (Special Operations) (Airborne), and Special Mission Units. The principal audience for Field Manual 3-05 is Army special operations forces, joint, and land component force commanders and staff; the publication provides a broad understanding of Army special operations forces. This publication also provides guidance for Army special operations forces commanders who determine the force structure, budget, training, materiel, and operations and sustainment requirements necessary to prepare Army special operations forces to conduct their missions. This Service doctrine is consistent with joint doctrine.

https://armypubs.us.army.mil/doctrine/DR_pubs/dr_c/pdf/fm3_05.pdf

FM 3-18 Special Forces Operations, 08 May 2014

FM 3-18 is the principal manual for Special Forces (SF) doctrine. It describes SF roles, missions, capabilities, organization, mission command, employment, and sustainment operations across the range of military operations. This manual is a continuation of the doctrine established in the JP 3-05 series, ADP 3-05, ADRP 3-05, and FM 3-05.

The principal audience for FM 3-18 is all members of the profession of arms. Commanders and staffs of Army headquarters serving as joint task force (JTF) or multinational headquarters should also refer to applicable joint or multinational doctrine concerning the range of military operations and joint or multinational forces. Trainers and educators throughout the Army will also use this publication.

https://armypubs.us.army.mil/doctrine/DR_pubs/dr_c/pdf/fm3_18.pdf

Army Training Publications (ATP)

ATP 3-05.1 Unconventional Warfare, September 2013
ATP 3-05.1, Unconventional Warfare, is the Army's doctrinal foundation for UW and is the broadest and most comprehensive United States Government (USG) doctrinal publication on the subject of UW. ATP 3-05.1 therefore provides doctrine directly useful to all users within the U.S. Army, but is deliberately intended to be useful to other Services in the Department of Defense (DOD) and joint, interagency, intergovernmental, and multinational (JIIM) audiences. Moreover, although UW is inherently a sensitive subject, ATP 3-05.1 is intentionally kept unclassified to make it accessible to civilian policy makers with a role in oversight and support of UW activities. ATP 3-05.1 is written to emphasize the strategic and operational utility of UW as a policy option available to national-level and theater-level decision makers. The ATP is therefore written for planners at the TSOC and SF group level who would be charged with recommending and planning strategic and operational options to geographic combatant commanders (GCCs), Ambassadors, and interagency decision makers at all levels of the USG. ATP 3-05.1 contains five chapters and six appendices that are summarized in the following paragraphs.

https://armypubs.us.army.mil/doctrine/DR_pubs/dr_c/pdf/atp3_05x1.pdf

ATP 3-05.1, C1 Unconventional Warfare, 25 November 2015
Army Techniques Publication (ATP) 3-05.1, Unconventional Warfare, provides the current United States (U.S.) Army Special Forces (SF) concept of planning and conducting unconventional warfare (UW) operations. ATP 3-05.1 describes UW fundamentals, activities, and considerations involved in the planning and execution of UW throughout the full range of military operations, and emphasize UW as a strategic policy option. This publication serves as the doctrinal foundation for subordinate Army special operations forces (ARSOF) UW doctrine, force integration, materiel acquisition, professional education, and individual and unit training. This publication also serves as the Army's description of UW, which will be useful in the larger joint and interagency environment.

https://armypubs.us.army.mil/doctrine/DR_pubs/dr_c/pdf/atp3_05x1c1.pdf

ATP 3-18.20, (S//NF) Advanced Special Operations Techniques (ASOT) (U), 30 December 2015
Defines the Special Forces concept of planning and employing ASOT in support of special operations forces (SOF) core activities, describes the fundamentals, activities, and considerations for planning and employing ASOT, and serves as the doctrinal foundation for ASOT to Army special operations forces (ARSOF). ATP 3-18.20 applies to ARSOF and may apply to

NAVSPECWARCOM and MARSOC Active, National Guard, and Reserve units, and is useful to the larger joint and interorganizational environment.
(SIPR ACCESS ONLY)

ATP 3-18.72 (S//NF) Special Forces Personnel Recovery (U), 13 January 2016
ATP 3-18.72 provides the doctrinal framework for U.S. Army special operations forces (ARSOF) personnel recovery operations from both the perspective of the recovery force, as well as that of the individual evader. It also provides an explanation of the various personnel recovery mission tasks, capabilities, limitations, general guidance, and employment techniques at both the strategic and tactical levels. Prepared under the direction of the Special Forces Doctrine Division, United States Army Special Operations Center of Excellence, USAJFKSWCS. This publication outlines the contributions of SF to the theater personnel recovery effort. SF personnel recovery missions seek to achieve specific, well defined and often sensitive results of strategic or operational significance. SF personnel recovery missions are conducted in support of their own operations , when directed by the joint task force commander to support combat search and rescue (CSAR) operation, when the threat to the recovery force is high enough to warrant the conduct of a special operation, and when SF are the only forces available or capable.
(SIPR ACCESS ONLY)

Training Circulars (TC)

TC 18-01 Special Forces Unconventional Warfare, January 2011
Training Circular (TC) 18-01, Special Forces Unconventional Warfare, defines the current United States (U.S.) Army Special Forces (SF) concept of planning and conducting unconventional warfare (UW) operations. For the foreseeable future, U.S. forces will predominantly engage in irregular warfare (IW) operations. TC 18-01 is authoritative but not directive. It serves as a guide and does not preclude SF units from developing their own standing operating procedures (SOPs) to meet their needs. It explains planning and the roles of SF, Military Information Support operations (MISO), and Civil Affairs (CA) in UW operations. There are appropriate manuals within the series that addresses the other primary SF missions in detail. The primary users of this manual are commanders, staff officers, and operational personnel at the team (Special Forces operational detachment A [SFODA]), company (Special Forces operational detachment B [SFODB]), and battalion (Special Forces operational detachment C [SFODC]) levels. This TC is specifically for SF Soldiers; however, it is also intended for use Army wide to improve the integration of SF into the plans and operations of other special operations forces (SOF) and conventional forces.
https://armypubs.us.army.mil/doctrine/DR_pubs/dr_c/pdf/tc18_01.pdf

TC 31-16 (S//NF) Special Forces Guide to Preparation of the Environment (U), 6 June 2007
This TC provides concepts, functions, and procedures for SF conducting PE to support current and future SOF operations.
(SIPRNET) *https://intelshare.intelink.sgov.gov/site/usasocg3/uwpedia*

Research and Science

USASOC Assessing Revolutionary and Insurgency Strategies (ARIS)
The Assessing Revolutionary and Insurgent Strategies (ARIS) project consists of research conducted for the US Army Special Operations Command by the National Security Analysis Department of The Johns Hopkins University Applied Physics Laboratory. Its goal is to produce academically rigorous yet operationally relevant research to develop and illustrate a common understanding of insurgency and revolution. Intended to form a bedrock body of knowledge for members of the Special Operations Forces, the ARIS studies allow users to distill vast amounts of material from a wide array of campaigns and extract relevant lessons, enabling the development of future doctrine, professional education, and training. The ARIS project follows in the tradition of research conducted by the Special Operations Research Office (SORO) of American University in the 1950s and 1960s, adding new research to that body of work, republishing original SORO studies, and releasing updated editions of selected SORO studies.
http://www.soc.mil/ARIS/ARIS.html

Casebook on Insurgency and Revolutionary Warfare, Volume I: 1933-1962, 25 January 2013 (Rev Ed.)
This casebook provides summary descriptive accounts of 23 revolutions that have occurred in seven geographic areas of the world, mostly since World War II. These areas include Southeast Asia, Latin America, North Africa, sub-Sahara Africa, Middle East, Far East and Europe. Each revolution is described in terms of the environment in which it occurred, the form of the revolutionary movement itself and the results it accomplished.
http://www.soc.mil/ARIS/CasebookV1S.pdf

Casebook on Insurgency and Revolutionary Warfare, Volume II: 1962-2009, 27 April 2012
This casebook provides summary descriptive accounts of 23 revolutions that have occurred since 1962. Many of these revolutions are still active-some in steady state violent conflict, others in decline, still others possibly approaching a complete resurgence. Each case presents a background of physical, cultural, social, economic and political factors that are relevant and important to understand the revolution.
http://www.soc.mil/ARIS/Casebook%20Vol%202%20%2004-27-12S.pdf

Undergrounds in Insurgency, Revolutionary, and Resistance Warfare, 25 January 2013 (2nd Ed.)
Since the original publication of Undergrounds in Insurgent, Revolutionary, and Resistance Warfare in 1963, much has changed yet much has remained the same. The internet, globalization,, social media, the demise of the Soviet Communism and the Cold War, and the rise of Islamic fundamentalism have all impacted the nature and functionality of undergrounds. The purpose of this book is to educate the student and practitioner of insurgency and counterinsurgency and to examine the anatomy of undergrounds in various insurgencies of recent history.
http://www.soc.mil/ARIS/UndergroundsS.pdf

Human Factors Considerations of Undergrounds in Insurgencies, 25 January 2013 (2nd Ed.)

During the 1950s through the mid-1960s social scientists and military personnel researched relevant political, cultural, social, and behavioral issues occurring within the emerging nations of Asia, Africa and Latin America. The Army had a particular interest in understanding the process of violent social change in order to be able to cope directly or indirectly or indirectly through assistance and advice with revolutionary actions. This book, Human Factors Considerations of Undergrounds in Insurgencies, is the second edition to the 1966 book of the same name and delves deeper into theory and further into background materials and focuses less on operational details.
http://www.soc.mil/ARIS/HumanFactorsS.pdf

Irregular Warfare Annotated Bibliography, 2 June 2011

The aim of the bibliography is to provide readers and offering of both a more traditional military perspective as well as perspectives from social scientist, including political scientist, sociologist, psychologist and anthropologists studying similar phenomena. The following sources include general history and analysis related to each core task: operational or "how to" guides; and works discussing particular insurgencies.
http://www.soc.mil/ARIS/IWAnnotated_BibliographyS.pdf

Legal Implications of the Status of Persons in Resistance

Our nation requires a special warfare capability. That capability requires intellectual investment in evolving our understanding of the legal environment and how that environment impacts US policy options and potential UW campaigns. As the legal analysis demonstrates, there will be some cases in which both the person's status as well as US government and international policy toward a resistance movement and its activities will be vague at best. This finding reinforces that strategists and practitioners must anticipate ambiguity in UW campaigns. Readers are encouraged read, analyze, debate, challenge, and consider how this analysis could impact Special Forces' ability to perform its UW mission.
http://www.soc.mil/ARIS/ARIS_Legal_Status-BOOK.pdf

Case Studies in Insurgency and Revolutionary Warfare-Columbia (1964-2009)

This case study presents a detailed account of revolutionary and insurgent activities in Colom via during the period from 1964 until 2009. It is specifically intended to provide a foundation for Special Force personnel to understand the circumstances, environment, and catalyst for revolution; the organization of resistance or insurgent organizations and their development, modes of operation, external support, and successes and failures; the counterinsurgents' organization, modes of operation, and external support, as well as their effects on resistance; and the outcomes and long-term ramification of the revolutionary/ insurgent activities.
http://www.soc.mil/ARIS/ARIS_Colombia-BOOK.pdf

Case Study in Guerilla War: Greece During World War II, 1961 (Rev Ed.)

Greece was selected as a logical subject for a pilot study on a guerrilla campaign by this Office for a number of reasons. Many similarities and cogent analogies exist between the guerrilla war in Greece in the early 1940's and those conflicts which have since broken out in other areas. On the other hand, certain aspects of the Greek situation are unique and of specific value. A study of the guerrilla warfare in Greece provides extremely useful insights into various perplexing problems concerning the exploitation and countering of guerrilla groups.

http://www.soc.mil/ARIS/ARIS_Greece-BOOK-small.pdf

Casebook on Insurgency and Revolutionary Warfare: Algeria 1954-1962, 1963 (Rev Ed.)

The Algerian Revolution case study is not a chronological narrative of the revolution from beginning to end. That type of historical case study is valuable for many purposes and a number have been published (see Bibliography). Rather, this study attempts to analyze, individually and successively through time, a number of factors in the revolutionary situation and the revolutionary movement itself which, on the basis of prior studies of revolutions, have been identified as being generally related to the occurrence, form, and outcome of a revolution. The case study, then, is devised to test the "explanatory power" of certain statements of relationships in terms of their applicability to the Algerian Revolution in particular. For this reason the reader is urged to read the definition of terms and the conceptual framework underlying the study which appears in the Technical Appendix.

http://www.soc.mil/ARIS/ARIS_Algeria-BOOK-small.pdf

Casebook on Insurgency and Revolutionary Warfare: Cuba 1953-1959 1963 (Rev Ed.)

This report on the recent Cuban Revolution is the first in a series of studies analyzing the rapid, often violent, change in political and socioeconomic order which is usually called revolution. From the many instances of modern revolution, the Cuban revolution was selected for these reasons: Cuban interests and those of the West are closely related and have been so for a long time. Second, events subsequent to the revolution have shown that the action or the lack of action by powers outside Cuba had a profound effect on the outcome of the revolution. Third, because the external powers involved in the post-revolutionary situation in Cuba are Communist, there is concern that the Cuban Revolution becomes a prototype for Communist revolutions elsewhere in Latin America. Lastly, the final form of the Cuban Revolution was different from its initial manifestations. It appeared originally to be a political protest movement with moderate aims; it grew into a major upheaval which changed the foundations of Cuban life.

http://www.soc.mil/ARIS/ARIS_Cuba-BOOK-small.pdf

Casebook on Insurgency and Revolutionary Warfare: Guatemala 1944-1954, November 1964 (Rev Ed.)

The Guatemalan study examines and attempts to analyze the rise and demise of the Communist Party in relation to the political activities of Guatemalan military officers during a period between two revolutions: the 1944 revolution which brought to power a liberal government within which the Communist Party gained power; and the 1954 revolution which made a conservative military officer head of state. The study also examines economic, social, and political factors which have been identified as being generally related to the rise of communism in Guatemala.

http://www.soc.mil/ARIS/ARIS_Guatemala-BOOK-small.pdf

White Papers, Academic Studies, and Other Articles

UW Library – Reference Repository
Aggregation of over 1,000 indexed and searchable UW documents, including: resistance movements, insurgency resource academic studies, historical reports, theses, research, and analyses of various subjects and fields of study related to this aspect of human conflict.
(SIPRNET) *https://intelshare.intelink.sgov.gov/site/usasocg3/uwpedia*

Little Green Men: A Primer on Modern Russian Unconventional Warfare, Ukraine 2013–2014, by Robert R. Leonhard, Stephen P. Phillips, and the ARIS Team
This document is intended as a primer—a brief, informative treatment—concerning the ongoing conflict in Ukraine. It is an unclassified expansion of an earlier classified version that drew from numerous classified and unclassified sources, including key US Department of State diplomatic cables.
http://www.jhuapl.edu/ourwork/nsa/papers/ARIS_LittleGreenMen.pdf

Russian UW in the Ukraine, USASOC, 2014 (S//NF)
Vladimir Putin spoke at a collegium of the ministry of defense where he stated "Our task – to create mobile, well-equipped armed forces, ready to promptly and adequately respond to potential threats."
(SIPRNET) *https://intelshare.intelink.sgov.gov/site/usasocg3/uwpedia*

Hybrid Structures, USASOC, 26 September 2014
This paper is intended to serve as a catalyst to generate discourse among the ARSOF and Conventional Force communities to explore new and innovative concepts, doctrine, partnerships and technological advancements to fulfill the ARSOF mission command capabilities required after 2022.
http://www.soc.mil/AUSA2014/Hybrid%20structures%20White%20Paper.pdf

Counter-Unconventional Warfare, 26 September 2014, USASOC
Hybrid Warfare involves a state or state-like actor's use of all available diplomatic, informational, military, and economic means to destabilize an adversary. Whole-of-government by nature, Hybrid Warfare as seen in the Russian and Iranian cases places a particular premium on unconventional warfare (UW).
https://info.publicintelligence.net/USASOC-CounterUnconventionalWarfare.pdf

Redefining the Win, 4 January 2015, USASOC
The Redefined Win Concept centers on proactive U.S. competition with State / Non-State Actors for the relative superiority over the physical, cognitive and moral security of key populations in the areas we choose to campaign.
https://usasoc.soc.mil/usasoc/g9/co/_layouts/WordViewer.aspx?id=/usasoc/g9/co/Shared%20Documents/01%20White%20Papers/USASOC%20White%20Paper%20-%20Redefining%20the%20Win%20-%20Final%20(V3%208)%20-%20RAW.docx&DefaultItemOpen=1

SOF Support to Political Warfare, 29 April 2015, USASOC
This white paper presents the concept of SOF Support to Political Warfare to leaders and policymakers as a dynamic means of achieving national security goals and objectives. Embracing the whole-of-government framework with significant targeted military contributions, Political Warfare enables America's leaders to undertake proactive strategic initiatives to shape environments, preempt conflicts, and significantly degrade adversaries' hybrid and asymmetric advantages.
https://usasoc.soc.mil/usasoc/g9/co/Shared%20Documents/Forms/Universal%20Columns.aspx?RootFolder=%2Fusasoc%2Fg9%2Fco%2FShared%20Documents%2F01%20White%20Papers%2FCurrent

Cognitive Joint Force Entry, 26 September 2014, USASOC
This paper describes the emerging idea of Cognitive Joint Force Entry. It provides an initial framework to consider how Inform and Influence Activities (IIA) can contribute to success in the shaping phase of campaigning. It also presents Cognitive Joint Force Entry as a vital capability in a critical domain working at strategic windows of opportunity where other instruments of national power may not or cannot function.
http://www.soc.mil/AUSA2014/Cognitive%20Joint%20Force%20Entry%20White%20Paper.pdf

Unconventional Options for the Defense of the Baltic States: The Swiss Approach by Jan Osburg, RAND, 2016
This RAND perspective examines how key concepts and elements of the decentralized resistance approach that was part of Swiss military strategy during the Cold War could also benefit the defense of the Baltic states.
http://www.rand.org/content/dam/rand/pubs/perspectives/PE100/PE179/RAND_PE179.pdf

Unconventional Warfare in the Gray Zone, Joseph L. Votel, Charles T. Cleveland, Charles T. Connett, and Will Irwin, National Defense University Press, 1 January 2016
As Operations Enduring Freedom and Iraqi Freedom have come to an end, this article examines the current and future conditions and implications for SOF. As the U.S. Armed Forces are faced with growing fiscal constraints, a smaller military and waning support for further large-scale deployment of troops, the homeland will continue to face increasing threats and our response will take place in a segment of the conflict continuum that some are calling the "Gray Zone," or an area that occupies space in the peace-conflict continuum. It is in here, the "Gray Zone", that SOF is the preeminent force of choice in such conditions. Special warfare tasks will increasingly rely on SOF's ability to build trust and confidence with our indigenous partners—host nation military and paramilitary forces, irregular resistance elements through either Unconventional Warfare (UW) or Foreign Internal Defense (FID) —to generate the requisite mass through indigenous forces, to engage our adversaries and thus eliminating the need for a large U.S. force presence. This paper further examines one key requirement of determining what success looks like in the "Gray Zone" and establishing meaningful criteria for measuring the effectiveness of such operations. "Unconventional Warfare in the Gray Zone" provides a different lens in which to evaluate what winning is, especially in places or situations where the U.S Government is unlikely to commit large military formations in decisive engagements against similarly armed foes. As this paper suggests, "winning" in the "Gray Zone" is not our traditional perception, but is best summed up as the

U.S. Government maintaining positional advantage, specifically our ability to influence partners, populations, and threats toward achievement of our regional or strategic objectives. In essence, it means retaining decision space, maximizing desirable strategic options, or simply denying an adversary a decisive positional advantage.
http://ndupress.ndu.edu/Media/News/NewsArticleView/tabid/7849/Article/643108/unc onventional-warfare-in-the-gray-zone.aspx

"The New Social Media Paradox: a Symbol of Self-Determination or a Boon for Big Brother?" Sara Smyth, International Journal of Cyber Criminology, (August 2, 2012). 2011 International Journal of Cyber Criminology (IJCC) ISSN: 0974 – 2891 January – June 2012, Vol. 6 (1): 924–950
Examines social media applications and tools impact on demand for political reform. Part II discusses the use of Facebook, Twitter, Internet, and mobile phones by protesters around the world in 2011; Part III discusses democratic and authoritarian government response; Part IV studies relevant policy concerns in American and Canadian legal contexts; and Part V notes implications for other online social networking sites.
http://ssrn.com/abstract=2122939

"A Social Movement Approach to Unconventional Warfare," Doowan Lee, Special Warfare Magazine, U.S. Army Special Warfare Center and School, July 2013
Delineates major components of social movement theory to inform UW planners on how to foment a resistance movement capable of garnering popular support as well as waging guerrilla warfare. Illustrates how the social movement approach can be operationalized for UW campaigns to enhance operational flexibility and strategic utility of UW by incorporating the logic of social mobilization and understanding of how to leverage existing social infrastructure. Examines UW-relevant lessons from the Arab Spring and other resistance movements.
https://static.dvidshub.net/media/pubs/pdf_12346.pdf

"Quick Reference Guide of Terms" in "Defining War," Jeffrey Hasler, Special Warfare, US Army John F. Kennedy Special Warfare Center and School, January 2011.
Mr. Hasler discusses the importance in using properly approved doctrinal terms and definitions to provide continuity, unity, and clarity amongst Soldiers and leaders of every echelon.
http://www.soc.mil/swcs/swmag/archive/SW2401/SW2401DefiningWar.html

"Continuity in the Chinese Mind for War," Jeffrey Hasler, Special Warfare, US Army John F. Kennedy Special Warfare Center and School, July 2012.
This article asserts the People's Republic of China (PRC) as an expanding power and examines Chinese history to consider the continuity of special-warfare stratagem and will alive in the eternal Chinese military mind. It characterizes Chinese military tradition and discusses Confucianism, Lao Tzu's Tao Te Ching, and Sun Tzu's The Art of War, recommending further study of the PRC, Chinese traditions, and their challenges to American interests.

http://www.soc.mil/SWCS/SWmag/archive/SW2503/SW2503ContinuityInTheChineseMindForWar.html

"Crossing the Red Line: Social Media and Social Network Analysis for Unconventional Campaign Planning," Seth Lucente and Greg Wilson, Special Warfare Magazine, Special Operations Command, July 2013
Analysis of considerations in Syria to develop unconventional intervention strategies that achieve U.S. policy objectives and limit expenditures. Discusses creation of a common operational picture from which policymakers and SOF military commanders can make informed decisions using open source, social media, temporal records, geospatial data and relational analysis.
https://static.dvidshub.net/media/pubs/pdf_12346.pdf

Demystifying the Title 10-50 Debate: Distinguishing Military Operations, Intelligence Activities and Covert Action, Presidents and Fellows of Harvard College, Andru E. Wall, 2011
This article asserts that modern warfare requires close integration of military and intelligence forces. The Secretary of Defense possesses authorities under Title 10 and Title 50 and is best suited to lead U.S. government operations against external unconventional and cyber threats. Titles 10 and 50 create mutually supporting, not mutually exclusive, authorities. Operations conducted under military command and control pursuant to a Secretary of Defense-issued execute order are military operations and not intelligence activities. Attempts by congressional overseers to redefine military preparatory operations as intelligence activities are legally and historically unsupportable. Congress should revise its antiquated oversight structure to reflect our integrated and interconnected world.
http://www.soc.mil/528th/PDFs/Title10Title50.pdf

Guidelines for Relations Between U.S. Armed Forces and Non-Governmental Humanitarian Organizations in Hostile or Potentially Hostile Environments, 2007
Short pamphlet with recommended guidelines to facilitate interaction between U.S. Armed Forces and NGOs.
http://www.usip.org/sites/default/files/guidelines_pamphlet.pdf

Guide to Non-Governmental Organizations for the Military: A Primer, 2009
This book is about NGOs, often referred to as private voluntary organizations (PVOs), nonprofits, charities, and (humanitarian) aid organizations. Its aim is to orient the military with NGOs: their operations, strengths, limitations, budgets, practices, and other characteristics that make them unique actors in a large and dynamic humanitarian community. This book has been produced specifically for uniformed services personnel, but it may prove useful to others in understanding some of the military-specific issues in foreign aid. Designed as a quick reference, annex 2 is a compilation of the most informative websites highlighted in different sections of the book. Annex 1 covers the basics for NGOs commonly found in humanitarian emergencies around the world and ones the military are likely to encounter.
https://www.researchgate.net/publication/235163824_Guide_to_Nongovernmental_Organizations_for_the_Military_A_primer_for_the_military_about_private_voluntary_and_nongovernmental_organizations_operating_in_humanitarian_emergencies_globally **or**
http://fas.org/irp/doddir/dod/ngo-guide.pdf

31

Christophe Fournier, Doctors Without Borders, NATO Speech, December 8, 2009
Fournier explains why Doctors without Borders can never be part of a "military-humanitarian coalition", the importance to make a clear distinction between impartial humanitarian actors and other more partisan aid actors, and finally the harmful consequences on the local population when this distinction is blurred. This conference was an opportunity to clarify that Doctors without Borders doesn't believe in a unity of purpose, but in a mutual understanding with all warring parties that allows for the deployment of impartial aid operations in order to contain the devastations of war.
http://www.msf.org/article/nato-speech-christophe-fournier

(S//NF) Non Standard Logistics Support to Unconventional Warfare: Sourcebook for Planning and Capability Development (U), Matthew Boyer, Dwayne Butler, John Halliday, Kristan Kinghoffer, and Roy Speaks, RAND Corporation, February 2012
Examines the role and ability of USASOC logistics to support UW campaigns, with emphasis on those phases that require clandestine support to an insurgency. This study is of interest to military logistics planners and developers, particularly those supporting UW.
(SIPRNET) *https://intelshare.intelink.sgov.gov/sites/rand-cooperation*

"Personnel Recovery Operations for Special Operations Forces in Urban Environments: Modeling Successful Overt and Clandestine Methods of Recovery," Marshall Ecklund and Michael McNerney, June 2004
This thesis presents two prescriptive models for approaching challenges to SOF with regard to PR in an urban environment. It begins by developing a model for overt recovery methods using McRaven's model of Special Operations as the foundation. This model is then tested against three different case studies from operations in Mogadishu, Somalia in 1993. The original six principles proposed by McRaven are complimented with four newly prescribed principles that account for the interactions of the isolated personnel. A nonconventional assisted recovery (NAR) model is presented for clandestine personnel recovery methods. This model borrows the relative superiority concept from McRaven's theory, but proposes six different principles. The model is evaluated using three case studies, one from the French theater of operations during World War II, another from the Korean War, and the third from Operation Iraqi Freedom.
http://www.dtic.mil/docs/citations/ADA425043

"The US Personnel Recovery Architecture under Chief of Mission Responsibility: Department of State and Department of Defense Coordination," Alejandro Nunez 22 May 2013
This paper discusses the importance of identifying gaps in PR architectures, determining how the DoS and DoD effectively coordinate during a PR event (within a Chief of Mission environment), and focuses on U.S. Army operations in the Western Hemisphere.
http://www.dtic.mil/docs/citations/ADA590491

"Whole of Government Approach to Personnel Recovery", William J. Rowell, 22 Mar 2012
This paper aims to create a shared understanding of the specific and even unique aspects of personnel recovery at the strategic level. An examination of Annex 1 to National Security

32

Presidential Directive - 12 provides a policy understanding that incorporates personnel recovery into a holistic government approach. This paper describes personnel recovery architecture, the two fundamental models used overseas, and recommends the development of a national strategy for PR.
http://www.dtic.mil/docs/citations/ADA561846

Books

iGuerilla: Reshaping the Face of War in the 21st Century, 20 May 2015, John Sutherland
ISBN-10: 1940773105
ISBN-13: 978-1940773100
Author John Sutherland alerts those in the United States and Western Europe to the threat posed by the modern Salafist-Jihadi insurgency which has infiltrated the West via virtual and physical means. Sutherland argues that the Wiki-warrior phenomena of armed insurgents empowered by globalization, information-based technologies, and sophisticated network methodologies are not an aberration, and represent a grave threat to international security.

Invisible Armies: An Epic History of Guerrilla Warfare from Ancient Times to the Present, 2013, Max Boot
ISBN-13: 978-0871406880
ISBN-10: 0871406888
Invisible Armies presents an entirely original narrative of warfare, which demonstrates that, far from the exception, loosely organized partisan or guerrilla are the historical norm.

U.S. Army Special Warfare: Its Origins: Revised Edition, 24 May 2002, Alfred H. Paddock Jr.
ISBN-10: 0700611770
ISBN-13: 978-0700611775
Special warfare was a key component of American military operations long before Afghanistan and even before the heroic deeds of the Green Berets. Alfred Paddock's revised edition of his classic study for two decades the definitive word on the subject honors the fiftieth anniversary of the organizations responsible for Army special warfare, and serves as a timely reminder of the likely role such forces can play in combating threats to American national security.

American Guerrilla: The Forgotten Heroics of Russell W. Volckmann, April 20, 2012, Mike Guardia
ISBN-10: 1612000894
ISBN-13: 978-1612000893
The Man Who Escaped from Bataan, Raised a Filipino Army against the Japanese, and became the True "Father" of Army Special Forces

Behind Japanese Lines: With the OSS in Burma, 4 February 2014, Richard Dunlop
ISBN-10: 1626365385
ISBN-13: 978-1626365384
In early 1942, with World War II going badly, President Roosevelt turned to General William "Wild Bill" Donovan, now known historically as the "Father of Central Intelligence," with orders

to form a special unit whose primary mission was to prepare for the eventual reopening of the Burma Road linking Burma and China by performing guerilla operations behind the Japanese lines.

The Jedburghs: The Secret History of the Allied Special Forces, France 1944
Paperback – October 10, 2006, Will Erwin
ISBN-10: 1586484621
ISBN-13: 978-1586484620
The story of the Special Forces in World War II has never fully been told before. Information about them began to be declassified only in the 1980s. Known as the Jedburghs, these Special Forces were selected from members of the British, American, and Free French armies to be dropped in teams of three deep behind German lines

Disrupting Dark Networks (Structural Analysis in the Social Sciences), 10 January 2013, Sean F. Everton
ISBN-10: 1107606683
ISBN-13: 978-1107606685
This is the first book in which counterinsurgency theory and social network analysis are coupled. Disrupting Dark Networks focuses on how social network analysis can be used to craft strategies to track, destabilize, and disrupt covert and illegal networks. The book begins with an overview of the key terms and assumptions of social network analysis and various counterinsurgency strategies.

Social Network Analysis: Methods and Applications (Structural Analysis in the Social Sciences), 25 November 1994, Katherine Faust
ISBN-10: 0521387078
ISBN-13: 978-0521387071
Social network analysis, which focuses on relationships among social entities, is used widely in the social and behavioral sciences, as well as in economics, marketing, and industrial engineering. Social Network Analysis: Methods and Applications reviews and discusses methods for the analysis of social networks with a focus on applications of these methods to many substantive examples

Comparative Perspectives on Social Movements: Political Opportunities, Mobilizing Structures, and Cultural Framings (Cambridge Studies in Comparative Politics), 26 January 1996, Doug McAdam, John D. McCarthy, Mayer N. Zald
ISBN-10: 0521485169
ISBN-13: 978-0521485166
Social movements such as environmentalism, feminism, nationalism, and the anti-immigration movement figure prominently in the modern world. Comparative Perspectives on Social Movements examines social movements in a comparative perspective, focusing on the role of ideology and beliefs, mechanisms of mobilization, and how politics shapes the development and outcomes of movements.

Dynamics of Contention (Cambridge Studies in Contentious Politics), 10 September 2001, Doug McAdam, Sidney Tarrow, Charles Tilly
ISBN-10: 0521011876
ISBN-13: 978-0521011877

Dissatisfied with the compartmentalization of studies concerning strikes, wars, revolutions, social movements, and other forms of political struggle, McAdam, Tarrow, and Tilly identify causal mechanisms and processes that recur across a wide range of contentious politics. Critical of the static, single-actor models (including their own) that have prevailed in the field, they shift the focus of analysis to dynamic interaction.

SOE: The Special Operations Executive 1940-1946
1984, M.R.D. Foot
ISBN-0-563-20193-2
SOE, the Special Operations Executive was small but tough British secret service that was set up in July 1940. It was composed of 10,000 men and 3,200 women of whom a full third were secret agents. The unit exercised vast influence on the war all over the world. This book clearly explains how SOE was run, the caliber of its men and women, the tools they used, how and where they applied them and their successes and failures

The Secret Army: The Memoirs of General Bor-Komorowski, December 1, 2011, Tadeuz Bor-Komorowski
ISBN-978-1-84832-595-1
The tale of Bor-Komorowski and the Uprising is the story of a proud nation and their fight against enemies and betrayal by allies. Based on true encounters when Germany invaded Poland in 1939. Bor was the commander of units defending the Vistula River before being pushed eastwards by the German's fierce advance. Despite being surrounded be German forces, he escaped to Cracow. Although he planned to escape to the West, he was ordered to stay and start a resistance movement. In the summer of 1941, he was sent to Warsaw and by 1943 had been appointed lead of the Home Army. The Polish resistance carried out sabotage and vital intelligence for the allies but their main task was to prepare for an uprising when the Nazis were in retreat to help liberate the country. After receiving orders from the Polish Government-in-Exile to commence operations on 1 August 1944, the Resistance held out for sixty-three days against vastly superior armaments. It was compromised when the oft-promise counter-attack from the Russians failed to materialize.

Guiding Principles for Stabilization and Reconstruction, United States Institute for Peace, November 1, 2009.
ISBN 978-1-60127-033-7
Discusses American Peace-building in developing countries.
http://www.usip.org/sites/default/files/guiding_principles_full.pdf

Videos

Defiance (2008)
A historic war drama based on the real-life Bielski Partisans. During World War II (1939-1945), two Polish Jewish brothers join anti-Nazi freedom fighters in the Białowieża Forest, straddling the border of Belarus and Poland, where they eventually manage to create a make-shift town harboring over 1000 Eastern European Jews. Details the tactics used by the Bielski Partisans to evade and counter-attack anti-partisan police forces and the German military.

Michael Collins (1996)

Biopic of Michael Collins, Irish revolutionary, veteran of the Easter Rising (1916) against the British Army, and leader of the Irish Republican Army (IRA) during the Irish War of Independence (1919-1921). Details Collins' role in the development of the IRA's anti-British guerrilla tactics; his emergence as one of the era's preeminent Irish political leaders and a key architect of the 1921 Anglo-Irish Treaty; and his death during the Irish Civil War (1922-1923).

Farewell to the King (1989)

A U.S. military deserter takes refuge among Dayak tribesmen in Borneo during World War II (1939-1945). After being named king of the tribe, the deserter is approached by British forces who aim to stage a guerrilla resistance struggle against the Japanese, who have occupied Borneo. The deserter mobilizes his tribe in the anti-Japanese struggle.

Red Dawn (1984)

This Cold War classic tells the story of a rag-tag group of Colorado high school students who take to the hills to wage a guerrilla insurgency against the Soviet-Cuban-Nicaraguan military forces who have invaded the continental United States, initiating World War III. Fighting behind enemy lines, living off the land, and using stealth and maneuverability, the kids manage to severely disrupt a much larger conventional military's campaign of occupation.

The Dogs of War (1980)

James Shannon, an American Vietnam War veteran working as mercenary, is hired by a British corporation to assess the fictional resource-rich West African country of Zangaro in order to determine the stability of the regime and the country's foreign investment potential.

Lawrence of Arabia (1962)

Englishman who led Arab Bedouin fighters against Ottoman Turkish military forces during World War I (1914-1918). Fluent in Arabic and an expert in Bedouin and Muslim culture, Lawrence lived among the Bedouin and managed to unite various warring tribes, leading them in a guerrilla campaign against the Turkish army.

The Special Operations Executive 1940 - 1946

What was the foundation of the underground army that helped turn the balance of power during World War Two? The agents of the SOE demonstrated tremendous courage, and enjoyed many successes, in their guerrilla war against Hitler's forces. This is their amazing story.
http://www.bbc.co.uk/history/worldwars/wwtwo/soe_01.shtml

Terms

Area Command - In unconventional warfare, the irregular organizational structure established within an unconventional warfare operational area to command and control irregular forces advised by Army Special Forces. It denotes the resistance leadership that directs, controls, integrates, and supports all resistance activities in the area of operations (AO). (ATP 3-05.1)

Area Complex - a clandestine, dispersed network of facilities to support resistance activities. It is a "liberated zone" designed to achieve security, control, dispersion, and flexibility. To support resistance activities, an area complex must include a security system, base camps, communications, logistics, medical facilities, supply caches, training areas, and escape and recovery mechanisms. The area complex may consist of friendly villages or towns under guerrilla military or political control. (ATP 3-05.1)

Auxiliary - For the purpose of unconventional warfare, the support element of the irregular organization whose organization and operations are clandestine in nature and whose members do not openly indicate their sympathy or involvement with the irregular movement. (ADRP 3-05)

Cache - A source of subsistence and supplies, typically containing items such as food, water, medical items, and/or communications equipment, packaged to prevent damage from exposure and hidden in isolated locations by such methods as burial, concealment, and/or submersion, to support isolated personnel. (JP 3-50)

Clandestine Operation - An operation sponsored or conducted by governmental departments or agencies in such a way as to assure secrecy or concealment. A clandestine operation differs from a covert operation in that emphasis is placed on concealment of the operation rather than on concealment of identity of the sponsor. In special operations, an activity may be both covert and clandestine and may focus equally on operational considerations and intelligence-related activities. (JP 3-05.1)

Covert Operation - An operation that is so planned and executed as to conceal the identity of or permit plausible denial by the sponsor. (JP 3-05)

Denied Area - An area under enemy or unfriendly control in which friendly forces cannot expect to operate successfully within existing operational constraints and force capabilities. (JP 3-05)

Government-in-Exile - A government that has been displaced from its country, but remains recognized as the legitimate sovereign authority. (ATP 3-05.1)

Guerrilla - An irregular, predominantly indigenous member of a guerrilla force organized similar to military concepts and structure in order to conduct military and paramilitary operations in enemy-held, hostile, or denied territory. Although a guerrilla and guerrilla forces can exist independent of an insurgency, guerrillas normally operate in covert and overt resistance operations of an insurgency. (ATP 3-05.1)

Guerrilla Base - A temporary site where guerrilla installations, headquarters, and some guerrilla units are located. A guerrilla base is considered to be transitory and must be capable of rapid displacement by personnel within the base. (ATP 3-05.1)

Guerrilla Force - A group of irregular, predominantly indigenous personnel organized along military lines to conduct military and paramilitary operations in enemy-held, hostile, or denied territory. (JP 3-05)

Guerrilla Warfare - Military and paramilitary operations conducted in enemy-held or hostile territory by irregular, predominantly indigenous forces. See also unconventional warfare. (JP 3-05.1)

Insurgency - The organized use of subversion and violence by a group or movement that seeks to overthrow or force change of a governing authority. Insurgency can also refer to the group itself. (JP 3-24)

Intelligence Operations - The variety of intelligence and counterintelligence tasks that are carried out by various intelligence organizations and activities within the intelligence process. (JP 2-01)

Intelligence Preparation of the Battlespace (IPB) - The analytical methodologies employed by the Services or joint force component commands to reduce uncertainties concerning the enemy, environment, time, and terrain. Intelligence preparation of the battlespace supports the individual operations of the joint force component commands. See also joint intelligence preparation of the operational environment. (JP 2-01.3)

Joint Intelligence Preparation of the Operational Environment (JIPOE) - The analytical process used by joint intelligence organizations to produce intelligence estimates and other intelligence products in support of the joint force commander's decision making process. It is a continuous process that includes defining the operational environment; describing the impact of the operational environment; evaluating the adversary; and determining adversary courses of action. (JP 2-01.3)

Low-Visibility Operations - Sensitive operations wherein the political-military restrictions inherent in covert and clandestine operations are either not necessary or not feasible; actions are taken as required to limit exposure of those involved and/or their activities. Execution of these operations is undertaken with the knowledge that the action and/or sponsorship of the operation may preclude plausible denial by the initiating power. (JP 3-05.1)

Mission Support Site - A preselected area used as a temporary base or stopover point. The mission support site is used to increase the operational range within the joint special operations area. (ATP 3-05.1)

Nonconventional Assisted Recovery (NAR) - Personnel recovery conducted by indigenous/surrogate personnel that are trained, supported, and led by special operations forces, unconventional warfare ground and maritime forces, or other government agencies' personnel that have been specifically trained and directed to establish and operate indigenous or surrogate infrastructures. Also called NAR. (JP 3-50) (Also refer to DODI 3002.04, Nov 17, 2014.)

Operational Environment - A composite of the conditions, circumstances, and influences that affect the employment of capabilities and bear on the decisions of the commander. (JP 3-0)

Personnel Recovery (PR) - The aggregation of military, civil, and political efforts to recover captured, detained, evading, isolated or missing personnel from uncertain or hostile environments and denied areas. Personnel recovery may occur through military action, action by nongovernmental organizations, other U.S. Government-approved action, and diplomatic initiatives, or through any combination of these options. Although personnel recovery may occur during noncombatant evacuation operations (NEOs), NEO is not a subset of personnel recovery. (DODD 2310.2)

Pilot Team - A deliberately structured composite organization comprised of Special Forces operational detachment members, with likely augmentation by interagency or other skilled personnel, designed to infiltrate a designated area to conduct sensitive preparation of the environment activities and assess the potential to conduct unconventional warfare in support of U.S. objectives. (ATP 3-05.1)

Preparation of the Environment - An umbrella term for operations and activities conducted by selectively trained special operations forces to develop an environment for potential future special operations. (JP 3-05)

Recovery Mechanisms (RM) - Indigenous or surrogate infrastructure that is specifically developed, trained, and directed by United States forces to contact, authenticate, support, move, and exfiltrate isolated personnel from uncertain or hostile areas back to friendly control. Also called RM. (JP 3-50)

Recovery Teams (RT) – In personnel recovery, designated United States or United States-directed forces, that are specifically trained to operate in conjunction with indigenous or surrogate forces, and are tasked to contact, authenticate, support, move, and exfiltrate isolated personnel. Also called RT. (JP 3-50)

Resistance Movement - An organized effort by some portion of the civil population of a country to resist the legally established government or an occupying power and to disrupt civil order and stability. (JP 3-05)

Shadow Government - Governmental elements and activities performed by the irregular organization that will eventually take the place of the existing government.

Members of the shadow government can be in any element of the irregular organization (underground, auxiliary, or guerrilla force). (ATP 3-05.1)

Subversion - Actions designed to undermine the military, economic, psychological, or political strength or morale of a governing authority. See also unconventional warfare. (JP 3-24)

Subversive Political Action - A planned series of activities designed to accomplish political objectives by influencing, dominating, or displacing individuals or groups who are so placed as to affect the decisions and actions of another government. (JP 1-02)

Unconventional Assisted Recovery (UAR) - Nonconventional Assisted Recovery conducted by special operations forces. (JP 3-50)

Unconventional Assisted Recovery Coordination Cell (UARCC) - A compartmented special operations forces facility, established by the joint force special operations component commander, staffed on a continuous basis by supervisory personnel and tactical planners to coordinate, synchronize, and de-conflict nonconventional assisted recovery operations within the operational area assigned to the joint force commander. (JP 3-50)

Unconventional Assisted Recovery Mechanism (UARM) - UARM encompasses SOF activities related to the creation, coordination, supervision, command and control, and use of recovery mechanisms either in support of Combatant Commands, or as directed by the National Command Authorities. UARM's may involve using an unconventional assisted recovery team (DODI 2310.6)

Unconventional Assisted Recovery Team (UART) - A designated special operations forces unconventional warfare ground or maritime force capable of conducting unconventional assisted recovery with indigenous or surrogate forces. (DODI 2310.6)

Underground - A cellular covert element within unconventional warfare that is compartmentalized and conducts covert or clandestine activities in areas normally denied to the auxiliary and the guerrilla force. (ADRP 3-05)

www.ingramcontent.com/pod-product-compliance
Lightning Source LLC
Chambersburg PA
CBHW070032030426
42335CB00017B/2400